The Illustrated Method of Archimedes:

Utilizing the Law of the Lever to Calculate Areas, Volumes, and Centers of Gravity

Andre K. T. Assis and C. P. Magnaghi

Apeiron
Montreal

Published by C. Roy Keys Inc.
4405, rue St-Dominique
Montreal, Quebec H2W 2B2 Canada
http://redshift.vif.com

First Published 2012

Library and Archives Canada Cataloguing in Publication

Assis, André Koch Torres, 1962-
The illustrated method of Archimedes : utilizing the law of the lever to calculate areas, volumes, and centers of gravity / Andre K.T. Assis and C.P. Magnaghi.

Includes bibliographical references.
ISBN 978-0-9864926-7-9

1. Mechanics. 2. Archimedes. I. Magnaghi, C. P. (Ceno Pietro), 1942- II. Title.

QA805.A87 2012 531 C2012-900753-6

Front cover: Illustration of the method of Archimedes. Figures in equilibrium along the arms of a lever.

Contents

Acknowledgments

The authors wish to thank Fernando Jorge da Paixão Filho, Sandro Guedes de Oliveira, Varlei Rodrigues, Márcio A. A. Pudenzi, Adolfo Maia Jr., Domingos S. d. L. Soares, Daniel Robson Pinto and João Paulo Martins de Castro Chaib for comments, suggestions, references, figures, etc.

Andre Koch Torres Assis and Ceno Pietro Magnaghi
Institute of Physics
University of Campinas—UNICAMP
13083-859 Campinas - SP, Brazil
E-mails: assis@ifi.unicamp.br and cenopietro@gmail.com
Homepage: www.ifi.unicamp.br/~assis

Chapter 1

Introduction

In 1906 Johan Ludwig Heiberg (1854-1928), Figure 1.1, a Danish philologist and historian of science, discovered a previously unknown text of Archimedes (287-212 B.C.).

Figure 1.1: J. L. Heiberg.

This work was known by different names: *Geometrical Solutions Derived from Mechanics*,[1] *The Method of Mechanical Theorems*,[2] and *The Method of Archimedes Treating of Mechanical Problems*.[3] Here it will simply be called *The Method*.

[1] [1] and [2].
[2] [3] and [4].
[3] [4].

This work was contained in a letter addressed to Eratosthenes (285-194 B.C.). In it, Archimedes presented a heuristic method to calculate areas, volumes and centers of gravity of geometric figures utilizing the law of the lever.

The goal of this work is to present the essence of Archimedes's method. The analysis included here will concentrate upon the physical aspects of these calculations. Figures will illustrate all levers in equilibrium. The postulates utilized by Archimedes will be emphasized. The mathematics will be kept to the minimum necessary for the proofs.

Chapter 2

The Physical Principles of Archimedes's Method

2.1 The Center of Gravity

2.1.1 Definition of the Center of Gravity

Archimedes mentioned the "center of gravity" of bodies many times in his memoirs . However, in his extant works there is no definition of this concept. It was probably included in a memoir that is now lost. In any event, from the analysis of his known works it appears that this concept might be understood as follows:[1]

> The center of gravity of any rigid body is a point such that, if the body be conceived to be suspended from that point, being released from rest and free to rotate in all directions around this point, the body so suspended will remain at rest and preserve its original position, no matter what the initial orientation of the body relative to the ground.

2.1.2 Experimental Determination of the Center of Gravity

From what has been found in his works, we can conclude that Archimedes knew how to experimentally determine the center of gravity of any rigid body. In Proposition 6 of his work *Quadrature of the Parabola*, he wrote:[2]

> Every suspended body — no matter what its point of suspension — assumes an equilibrium state when the point of suspension and

[1] [5, pp. 24, 301, 350-351 and 430], [3, pp. 17, 47-48, 289-304, 315-316, 321-322 and 435-436], [6, pp. clxxxi-clxxxii] and [7, Chapter 6, pp. 123-132].

[2] [8, p. 463], [9, p. 307], [10, p. 171] and [7, p. 124].

the center of gravity are on the same vertical line. This has been demonstrated.

This suggests a practical procedure for finding the center of gravity of a body experimentally, as follows. Suspend the rigid body by a point of suspension P_1. Wait until the body reaches equilibrium and draw the vertical passing through the P_1 with the help of a plumb line. Let E_1 be the extremity of the body along this vertical line, Figure 2.1.

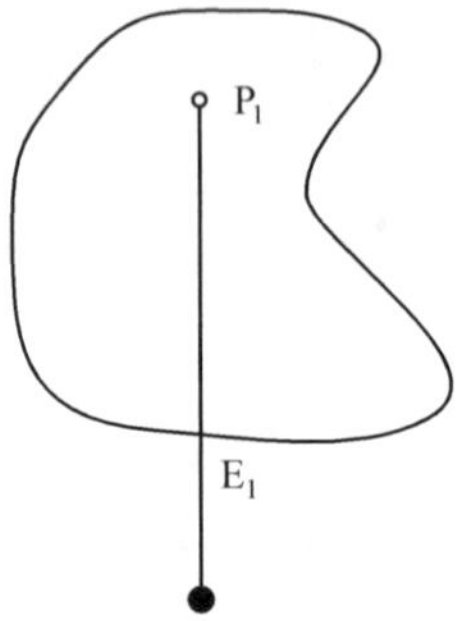

Figure 2.1: A plumb line is utilized to draw the vertical line connecting the suspension point P_1 to the extremity E_1 of the body.

Then suspend the body by another point of suspension P_2 which is not along the first vertical P_1E_1. Wait until it reaches equilibrium, and draw a second vertical through P_2, connecting it to the extremity E_2 of the body along this vertical. The intersection of the two verticals is the center of gravity of the body, CG, Figure 2.2.

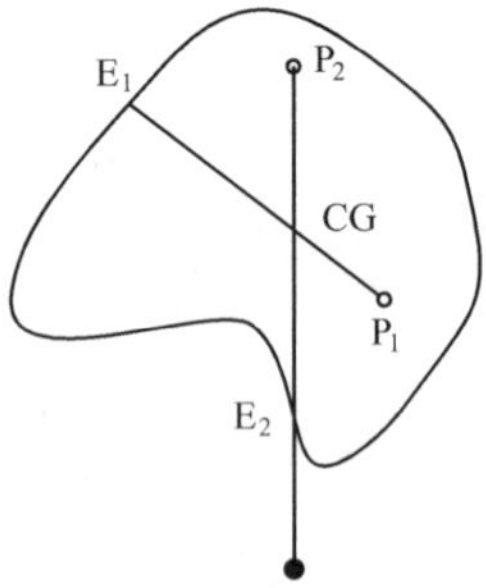

Figure 2.2: The intersection of two verticals is the center of gravity of the body.

It should be emphasized that, according to Archimedes, this was not a definition of the center of gravity. Instead, he proved this result theoretically utilizing a previous definition of the center of gravity of a body, as well as some postulates that are now unknown.

2.1.3 Theoretical Determination of the Center of Gravity

In his works Archimedes calculated the center of gravity of one, two and three-dimensional figures.[3]

One of his most important postulates which he utilized to arrive at these results was the famous sixth postulate of his work *On the Equilibrium of Planes.* It reads as follows:[4]

> If magnitudes at certain distances be in equilibrium, (other) magnitudes equal to them will also be in equilibrium at the same distances.

The meaning of this crucial sixth postulate has been clarified by Vailati, Toeplitz, Stein and Dijksterhuis.[5] By "magnitudes equal to other magnitudes," Archimedes wished to say "magnitudes of the same weight." And by "magnitudes at the same distances," Archimedes understood "magnitudes the centers of gravity of which lie at the same distances from the fulcrum."

Suppose a system of bodies maintains a lever in equilibrium. According to this postulate, a certain body A suspended from the lever can be replaced by another body B, without disturbing equilibrium, provided the following conditions are satisfied: (1) the weight of B must be equal to the weight of A; and (2) both bodies must be suspended by their centers of gravity which lie at the same distance from the fulcrum.

In his work *On the Equilibrium of Planes* Archimedes utilized this sixth postulate to demonstrate the law of the lever and to calculate the center of gravity of a triangle, etc.[6]

This sixth postulate is also essential in the method of Archimedes discussed in this work, Chapter 4.

2.2 The Law of the Lever

The lever is one of the simple machines studied in ancient Greece. It consists of a rigid body, normally linear, the beam, capable of turning around a fixed axis horizontal to the ground. This axis is called the fulcrum or point of suspension, PS, of the lever. This axis is orthogonal to the beam. The lever is like a balance, but now with the possibility of placing weights at different distances from the fulcrum. It will be supposed that the lever is symmetrical about the vertical plane passing through the fulcrum, with the beam horizontal and orthogonal to this vertical plane when there are no bodies supported by the lever.

[3] See [7, Section 6.2: Theoretical Values of Center of Gravity Obtained by Archimedes, pp. 132-136].

[4] [6, p. 190], [11, p. 287] and [7, pp. 210-211].

[5] [12], [11, pp. 289-304 and 321-322] and [7, Section 7.1: Archimedes's Proof of the Law of the Lever, pp. 209-215].

[6] A detailed discussion of this work can be found in [7, Section 10.7: Archimedes's Proof of the Law of the Lever and Calculation of the Center of Gravity of a Triangle, pp. 209-217].

A lever is in equilibrium when its beam remains at rest horizontally relative to the ground. The horizontal distance d between the point of suspension of a body upon the beam and the vertical plane passing through the fulcrum is called the arm of the lever. For brevity, sometimes we utilize the following expression: "distance between the body and the fulcrum." This should be understood as meaning the horizontal distance between the point of suspension of the body upon the beam and the vertical plane passing through the fulcrum. When the two arms of a lever are mentioned, these should be understood as the opposite sides in relation to the vertical plane passing through the fulcrum.

Archimedes demonstrated the law of the lever in Propositions 6 and 7 of his work *On the Equilibrium of Planes*. These proposition are:[7]

> Proposition 6: Commensurable magnitudes are in equilibrium at distances reciprocally proportional to the weights.
> Proposition 7: However, even if the magnitudes are incommensurable, they will be in equilibrium at distances reciprocally proportional to the magnitudes.

Heath combined these two propositions in his paraphrase of Archimedes's work:[8]

> Propositions 6, 7. Two magnitudes, whether commensurable [Prop. 6] or incommensurable [Prop. 7], balance at distances reciprocally proportional to the magnitudes.

Suppose weights W_A and W_B are located on two sides of a lever in equilibrium supported by their centers of gravity located at distances d_A and d_B from the fulcrum F, Figure 2.3.

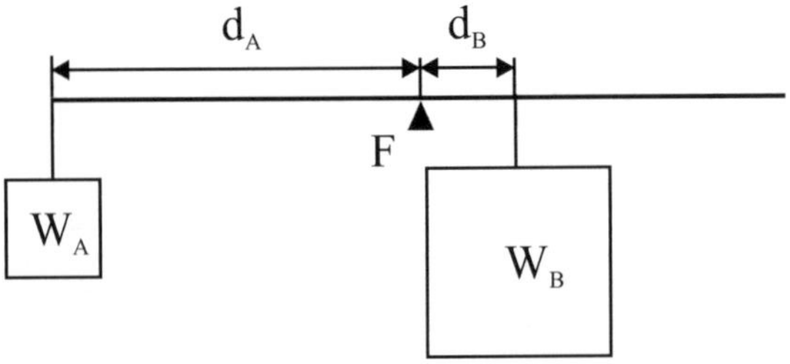

Figure 2.3: Lever in equilibrium about the fulcrum F.

According to the law of the lever, equilibrium will prevail if

$$\frac{d_A}{d_B} = \frac{W_B}{W_A} \ . \tag{2.1}$$

[7] [11, pp. 289 and 305].
[8] [6, p. 192].

Chapter 3

Archimedes, the Circle and the Sphere

Archimedes was greatly interested in the properties of the circle and the sphere. He knew that the circumference or perimeter of a circle is proportional to its diameter. In his time this theorem should be expressed as:

Circumferences are to one another as the diameters.

Let p_1 and p_2 be the circumferences or perimeters of circles with radii r_1 and r_2, respectively, as in Figure 3.1.

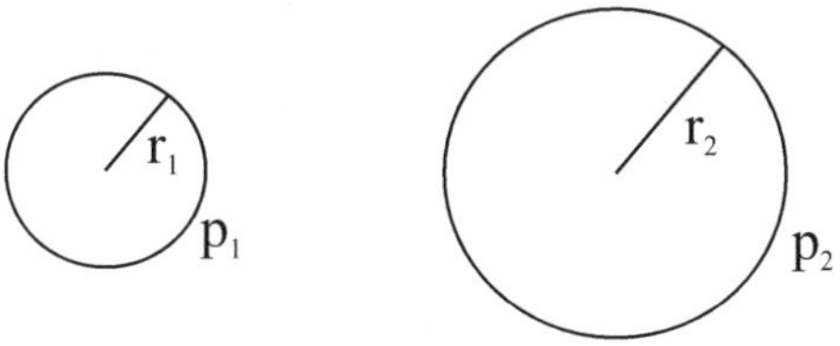

Figure 3.1: Circles with perimeters p_1 and p_2 and radii r_1 and r_2.

The previous theorem can be expressed mathematically as

$$\frac{p_1}{p_2} = \frac{2r_1}{2r_2} = \frac{r_1}{r_2} \ . \tag{3.1}$$

It was only in 1706 that the mathematician William Jones (1675-1749) proposed the use of the symbol π to represent the ratio of the circumference of a circle to its diameter. This definition of π was later popularized by the famous mathematician and physicist Leonhard Euler (1707-1783) in 1737. This definition can be expressed mathematically as follows, considering any circle of circumference p and radius r:

$$\pi \equiv \frac{p}{2r} \ . \tag{3.2}$$

With this definition the circumference of any circle can be written as:

$$p = 2\pi r \ . \tag{3.3}$$

It was only in 1761 that J. H. Lambert (1728-1777) proved that π is an irrational number, so that it cannot be written as the ratio of two integers.

Although Archimedes did not mention anything about the irrationality of the ratio of the circumference to the diameter of a circle, he obtained an excellent approximation for this constant ratio in his work *Measurement of a Circle.*[1] He found upper and lower bounds for this ratio by inscribing and circumscribing a circle with two similar n-sided regular polygons. Figure 3.2 shows a circle with inscribed and circumscribed hexagons. When the number n is increased, the perimeters of both polygons approach the circumference of the circle.

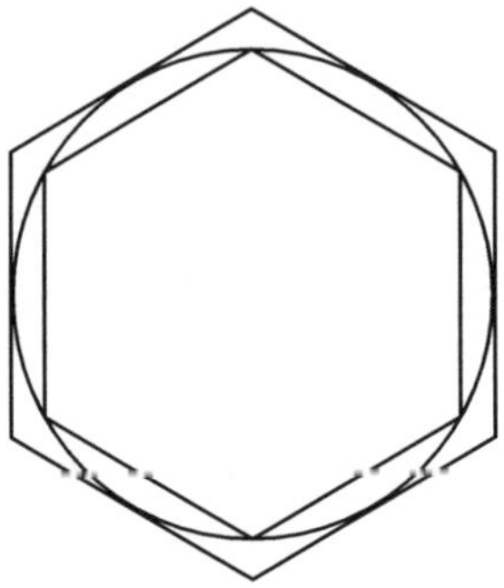

Figure 3.2: Circle with inscribed and circumscribed hexagons.

By going up to inscribed and circumscribed 96-sided regular polygons Archimedes found that:[2]

> The ratio of the circumference of any circle to its diameter is less than $3\frac{1}{7}$ but greater than $3\frac{10}{71}$.

Dijksterhuis expressed this theorem as follows:[3]

> The circumference of any circle is three times the diameter and exceeds it by less than one-seventh of the diameter and by more than ten-seventyoneths.

Mathematically this theorem can be expressed as follows:

$$3\frac{10}{71} < \frac{p}{2r} < 3\frac{1}{7} \ . \tag{3.4}$$

Equation (3.2) combined with Equation (3.4) yields:

[1] *Measurement of a Circle*, [6, pp. 91-98].
[2] [6, *Measurement of a Circle*, Proposition 3, p. 93].
[3] [3, *Measurement of the Circle*, Proposition 3, p. 223].

$$3.1408 < \pi < 3.1429 \ . \tag{3.5}$$

These were remarkable upper and lower bounds for the ratio of the circumference to the diameter of any circle expressed by means of such simple numbers.

Since Eudoxus and Euclid it was known that:[4]

> Circles are to one another as the squares on the diameters.

Consider the circles 1 and 2 of Figure 3.1 with areas $A_{Circle\ 1}$ and $A_{Circle\ 2}$, respectively. This theorem can be expressed mathematically as follows:

$$\frac{A_{Circle\ 1}}{A_{Circle\ 2}} = \left(\frac{2r_1}{2r_2}\right)^2 = \left(\frac{r_1}{r_2}\right)^2 \ . \tag{3.6}$$

Archimedes went a step further. In his work *Measurement of a Circle* he proved that:[5]

> The area of any circle is equal to a right-angled triangle in which one of the sides about the right angle is equal to the radius, and the other to the circumference, of the circle.

This is illustrated in Figure 3.3.

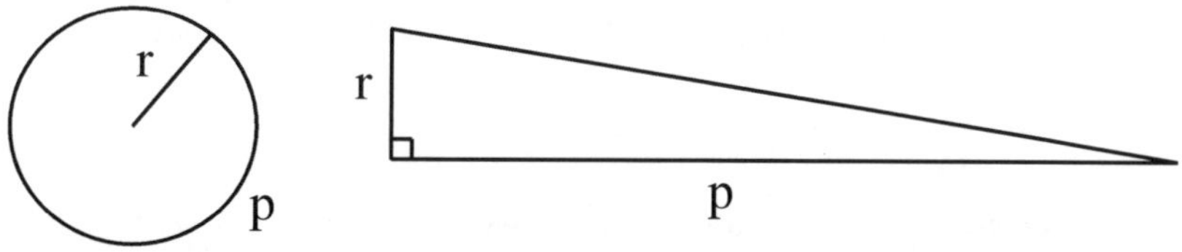

Figure 3.3: Archimedes proved that this circle and this triangle have the same area.

Let A_{Circle} be the area of a circle of radius r and circumference p. Let A_T be the area of the right-angled triangle in which one of the sides about the right angle is equal to the radius r, and the other side is equal to the circumference p, of the circle.

The result obtained by Archimedes in *Measurement of a Circle* is represented nowadays by the following formula:

$$A_{Circle} = A_T = \frac{p \cdot r}{2} \ . \tag{3.7}$$

Combining Equations (3.2), (3.3) and (3.7) yields the modern formula for the area of a circle, namely:

$$A_{Circle} = A_T = \frac{p \cdot r}{2} = \frac{2\pi r \cdot r}{2} = \pi r^2 \ . \tag{3.8}$$

[4] [13, Proposition 2, Book XII].

[5] *Measurement of a Circle*, Proposition 1, [6, p. 91].

Since Eudoxus and Euclid it was also known that:[6]

> Spheres are to one another in the triplicate ratio of their respective diameters.

Consider two spheres of volumes V_{S_1} and V_{S_2} and radii r_1 and r_2, respectively. This theorem can be expressed algebraically as follows:

$$\frac{V_{S_1}}{V_{S_2}} = \left(\frac{2r_1}{2r_2}\right)^3 = \left(\frac{r_1}{r_2}\right)^3 . \tag{3.9}$$

Archimedes went a step further. In the first part of his work On the Sphere and Cylinder he proved three extremely important results, namely:[7]

> Proposition 33: The surface of any sphere is equal to four times the greatest circle in it.
>
> Proposition 34: Any sphere is equal to four times the cone which has its base equal to the greatest circle in the sphere and its height equal to the radius of the sphere.
>
> Corollary. From what has been proved it follows that every cylinder whose base is the greatest circle in a sphere and whose height is equal to the diameter of the sphere is $\frac{3}{2}$ of the sphere, and its surface together with its bases is $\frac{3}{2}$ of the surface of the sphere.

Let A_S be the area of a sphere of radius r, with A_{Circle} being the area of the greatest circle in it. Proposition 33 can be expressed algebraically as follows:

$$A_S = 4A_{Circle} . \tag{3.10}$$

Equations (3.8) and (3.10) yield the modern result

$$A_S = 4A_{Circle} = 4\pi r^2 . \tag{3.11}$$

Proposition 34 can be represented by Figure 3.4.

Let V_S be the volume of a sphere of radius r and V_{Cone} the volume of a cone which has its base equal to the greatest circle of the sphere and its height equal to the radius of the sphere. This theorem can be expressed algebraically as follows:

$$V_S = 4V_{Cone} . \tag{3.12}$$

Since Democritus, Eudoxus and Euclid it was known that:[8]

> Any cone is a third part of the cylinder which has the same base with it and equal height.

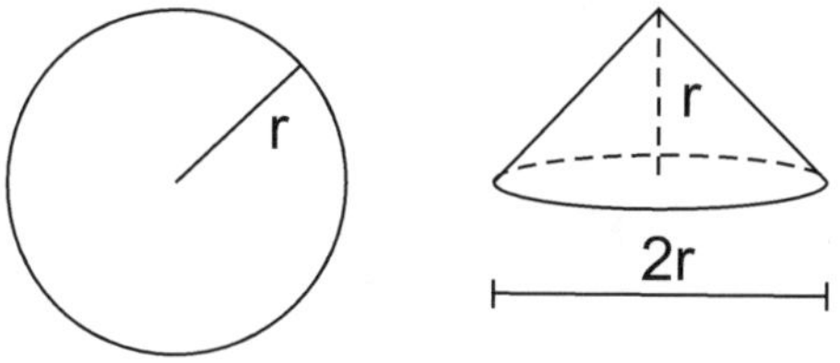

Figure 3.4: The volume of any sphere is equal to four times the volume of the cone which has its base equal to the greatest circle in the sphere and its height equal to the radius of the sphere.

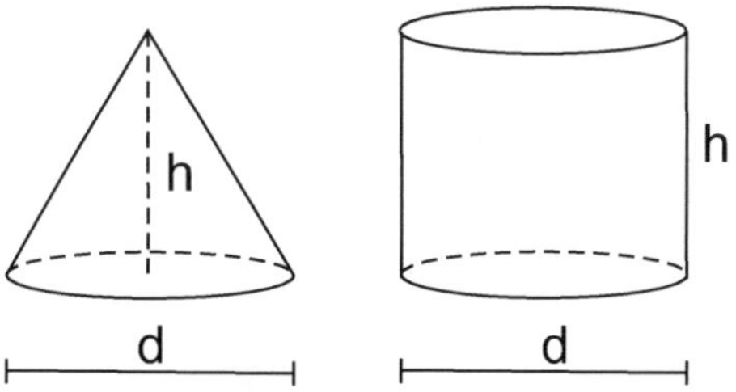

Figure 3.5: The volume of any cone is equal to a third of the volume of the cylinder which has the same base with it and equal height.

This theorem is illustrated by Figure 3.5.

Let V_{Cyl} be the volume of a cylinder and V_{Cone} the volume of a cone with the same height and equal base. This theorem can be expressed algebraically as follows:

$$V_{Cone} = \frac{1}{3} V_{Cyl} \ . \tag{3.13}$$

Equations (3.12) and (3.13) yield

$$V_S = 4V_{Cone} = \frac{4}{3} V_{Cyl} \ . \tag{3.14}$$

In Equation (3.14) the cone and the cylinder have height equal to the radius r of the sphere and base equal to the greatest circle of the sphere, Figure 3.6.

The volume of the cylinder of Equation (3.14) with height r is half the volume of the cylinder with height $2r$ circumscribing the sphere of radius r, as shown by Figure 3.7.

Let $V_{Circumscribing\ Cyl}$ be the volume of the cylinder circumscribing a sphere of radius r and volume V_S. Equation (3.14) can then be expressed as:

$$V_S = \frac{2}{3} V_{Circumscribing\ Cyl} \ . \tag{3.15}$$

[6][13, Proposition 18, Book XII].

[7][6, *On the Sphere and Cylinder*, Book I, Propositions 33 and 34, pp. 39-43].

[8][14, p. 13] and [13, Proposition 10, Book XII].

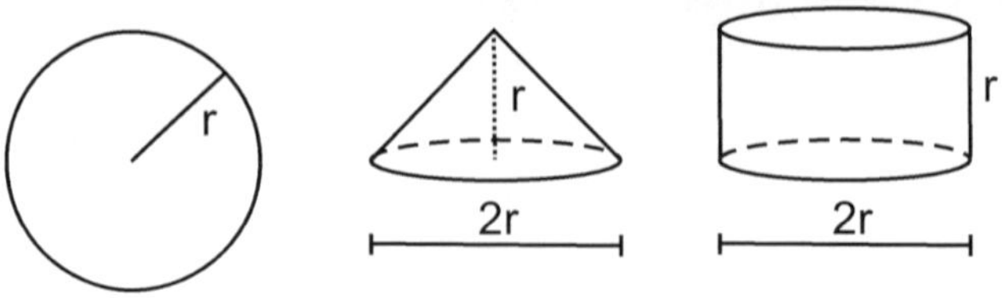

Figure 3.6: A sphere of radius r, a cone of height r and base equal to the greatest circle of the sphere, and a cylinder of height r and base equal that of the cone.

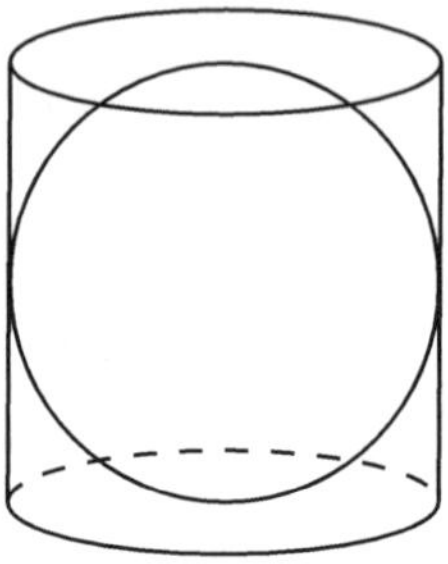

Figure 3.7: Sphere with circumscribing cylinder.

This is the result which Archimedes expressed in the Corollary to Proposition 34 of book I of his work *On the Sphere and Cylinder*, stating that the volume of the cylinder which has a great circle of the sphere for its base and whose height is equal to the diameter is equal to one and a half times the volume of the sphere, namely:

$$V_{Circumscribing\ Cyl} = \frac{3}{2} V_S \ . \tag{3.16}$$

But the volume of the circumscribing cylinder is equal to its base times its height. From Equation (3.8) its base is equal to πr^2 while its height is equal to $2r$, so that:

$$V_{Circumscribing\ Cyl} = \frac{3}{2} V_S = \left(\pi r^2\right) \cdot (2r) \ . \tag{3.17}$$

This Equation is analogous to the modern result, namely:

$$V_S = \frac{2}{3} V_{Circumscribing\ Cyl} = \frac{2}{3} \left(\pi r^2\right) \cdot (2r) = \frac{4}{3} \pi r^3 \ . \tag{3.18}$$

In essence, the modern results that the area of a circle of radius r is given by πr^2, the area of a sphere of radius r is given by $4\pi r^2$ and its volume is equal to $4\pi r^3/3$ are all due to Archimedes.

The proofs of Propositions 33 and 34 of his work *On the Sphere and Cylinder* were purely geometrical. It was only with the discovery of his work *The Method* that it became known how he first obtained these results. His heuristic method utilized the law of the lever. This will be shown in Section 4.3.

Chapter 4

The Illustrated Method of Archimedes

4.1 Lemmas of the Method

In the course of *The Method* Archimedes used several lemmas. The most relevant of these lemmas which are of interest here are presented below.[1]

> The center of gravity of any straight line is the point of bisection of the straight line.
>
> The center of gravity of any triangle is the point in which the straight lines drawn from the angular points of the triangle to the middle points of the (opposite) sides cut one another.
>
> The center of gravity of a circle is the point which is also the center [of the circle].
>
> The center of gravity of any cylinder is the point of bisection of the axis.

4.2 Physical Demonstration of Theorem I: Area of a Parabolic Segment

Figure 4.1 is a parabola $\rho\phi\gamma$ with vertex ϕ and diameter $\phi\eta$. This diameter is the axis of symmetry of the parabola. The chord $\rho\gamma$ is the base of the segment, being perpendicular to $\phi\eta$, where η is the middle point of $\rho\gamma$. The chord $\alpha\gamma$ is inclined relative to the diameter. The point δ divides $\alpha\gamma$ into two equal segments. From δ a straight line is drawn parallel to the diameter $\phi\eta$, meeting the parabola at β. Therefore $\beta\delta$ is parallel to $\phi\eta$.

In the particular case in which α coincides with ρ then the chord $\alpha\gamma$ will

[1] [4, pp. 14-15].

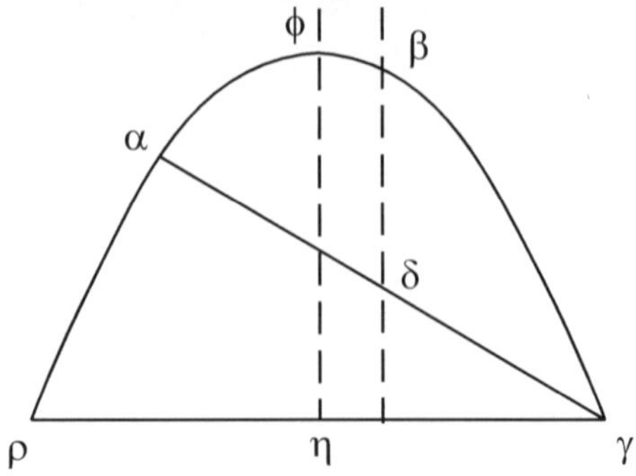

Figure 4.1: Parabola $\rho\phi\gamma$ with vertex ϕ, diameter $\phi\eta$ and chord $\rho\gamma$ perpendicular to the diameter, divided into two equal parts at η. Archimedes considered the general case of a parabolic segment $\alpha\beta\gamma$ with chord $\alpha\gamma$ inclined relative to the diameter $\phi\eta$. The point δ divides $\alpha\gamma$ into two equal segments while, by construction, $\beta\delta$ is parallel to $\phi\eta$.

coincide with $\rho\gamma$. In this case $\alpha\gamma$ will be perpendicular to the diameter $\phi\eta$, because β will coincide with ϕ, while δ will coincide with η, Figure 4.2 (a).

Archimedes considered the general case of the parabolic segment $\alpha\beta\gamma$ with chord $\alpha\gamma$ inclined relative to the diameter $\phi\eta$, as in Figure 4.2 (b). The symmetrical case occurs when α coincides with ρ, while β coincides with ϕ. This is a particular case contained in the general case.

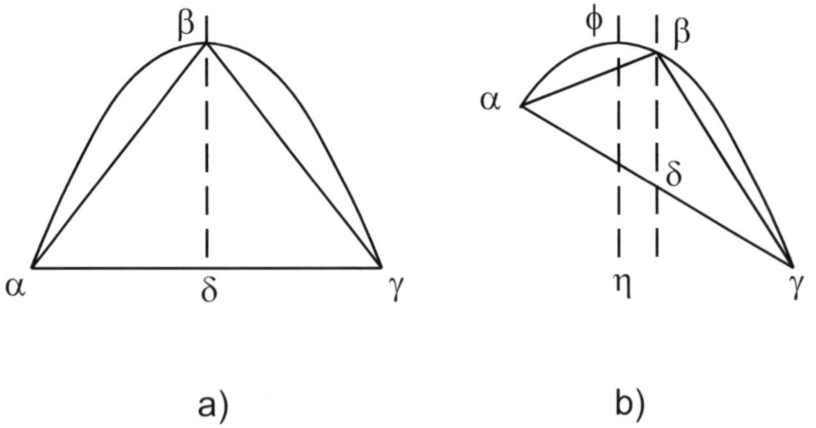

Figure 4.2: (a) Chord $\alpha\gamma$ perpendicular to the diameter $\phi\eta$, which coincides with $\beta\delta$. (b) Chord $\alpha\gamma$ inclined relative to the diameter $\phi\eta$.

Archimedes demonstrated that the area of the parabolic segment $\alpha\beta\gamma$ is equal to 4/3 the area of the triangle $\alpha\beta\gamma$ inscribed into this parabola. This result is valid not only in the symmetrical case for which the chord $\alpha\gamma$ is perpendicular to the diameter $\phi\eta$, Figure 4.2 (a), but also in the general case for which the chord $\alpha\gamma$ may be inclined relative to the diameter $\phi\eta$, Figure 4.2 (b).

That is, the following relation is valid in both cases:

$$\frac{\text{area of the parabolic segment } \alpha\beta\gamma}{\text{area of the triangle } \alpha\beta\gamma} = \frac{4}{3} . \tag{4.1}$$

Figure 4.3 presents the main elements to prove this Theorem.[2]

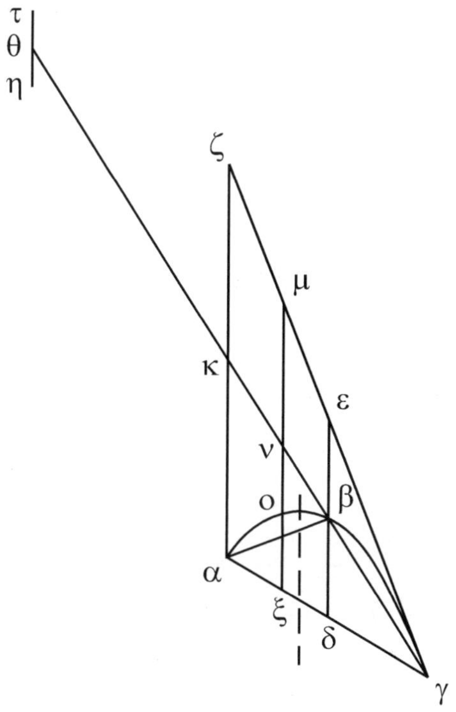

Figure 4.3: Geometric construction of Theorem I in the general case. The dashed straight segment is the diameter or axis of symmetry of the parabola.

In Figure 4.3 let $\alpha\beta\gamma$ be a segment of a parabola bounded by the straight line $\alpha\gamma$ and the parabola $\alpha\beta\gamma$, and let δ be the middle point of $\alpha\gamma$. From δ draw the straight line $\delta\beta\varepsilon$ parallel to the diameter of the parabola and join $\alpha\beta$ and $\beta\gamma$. From α draw $\alpha\kappa\zeta$ parallel to $\varepsilon\delta$, and let the tangent to the parabola at γ meet $\delta\beta\varepsilon$ in ε and $\alpha\kappa\zeta$ in ζ. Produce $\gamma\beta$ to meet $\alpha\zeta$ at κ, and again produce $\gamma\kappa$ to θ, making $\theta\kappa$ equal to $\kappa\gamma$. Consider the straight line $\xi\mu$ parallel to the diameter of the parabola and at an arbitrary distance from $\alpha\zeta$. The point o is the intersection of $\xi\mu$ with the parabola $\alpha\beta\gamma$, while the point ν is the intersection of $\xi\mu$ and $\kappa\gamma$. Archimedes showed in Proposition 2 of *Quadrature of the Parabola* that β is the middle point of $\varepsilon\delta$.[3] Then, by similarity of triangles,

[2] [11, p. 317] and [4, p. 16].

[3] [6, p. 235].

it can be shown that κ and ν are the middle points of the straight lines $\alpha\zeta$ and $\xi\mu$, respectively.

From the geometry of Figure 4.3, Archimedes proved the following result in *The Method*:[4]

$$\frac{\theta\kappa}{\kappa\nu} = \frac{\mu\xi}{o\xi} . \tag{4.2}$$

Archimedes then considered the segments $\mu\xi$ and ξo as having weights proportional to their lengths. He considered $\theta\gamma$ as the bar of a lever, κ being its fulcrum, the middle point of the lever. Take a straight line $\tau\eta$ equal to $o\xi$, and place it with its center of gravity at θ, so that $\tau\theta = \theta\eta$. The point ν is the center of gravity of the straight line $\mu\xi$, while θ is the center of gravity of $\tau\eta$. The law of the lever, Equation (2.1), combined with Equation (4.2), leads to the result that this lever $\theta\gamma$ will remain in equilibrium about κ with the segment $\mu\xi$ at ν and the segment $o\xi$ or $\tau\eta$ at θ. This equilibrium is represented in Figure 4.4.

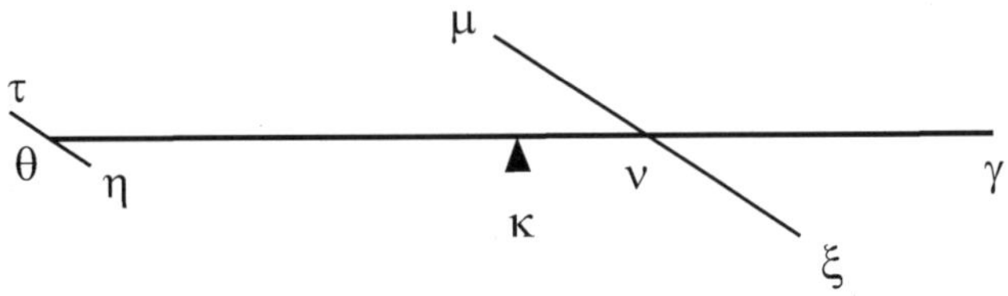

Figure 4.4: Equilibrium of the straight segments upon an horizontal lever.

This equilibrium is represented mathematically by the following equation:

$$\frac{\theta\kappa}{\kappa\nu} = \frac{\mu\xi}{\tau\eta} . \tag{4.3}$$

Similarly, for all other straight lines parallel to $\delta\varepsilon$ and meeting the arc of the parabola, (1) the portion intersected between $\zeta\gamma$ and $\alpha\gamma$ with its middle point on $\kappa\gamma$ and (2) a length equal to the intersect between the arc of the parabola and $\alpha\gamma$ placed with its center of gravity at θ will be in equilibrium about κ. The straight segments $o\xi$ from $\alpha\xi = 0$ up to $\alpha\xi = \alpha\gamma$ fill up the parabolic segment $\alpha\beta\gamma$ acting only at θ. The straight segments $\mu\xi$ fill up the triangle $\alpha\zeta\gamma$ distributed along $\kappa\gamma$. Therefore, a lever is obtained which is in equilibrium when supported at κ, with the segment of parabola $\alpha\beta\gamma$ acting by its center of gravity at θ, and the triangle $\alpha\zeta\gamma$ distributed along the arm $\kappa\gamma$. This is represented in Figure 4.5 with the segment of parabola suspended at θ by a weightless string, with its center of gravity vertically below θ.

By the sixth postulate of *On the Equilibrium of Planes* quoted in Subsection 2.1.3, this lever will remain in equilibrium about κ suspending by a weightless string the triangle $\alpha\zeta\gamma$ only by its center of gravity. That is, instead of being distributed along the arm of the lever, it will be suspended by a single point coinciding with its center of gravity.

[4] [4, p. 16].

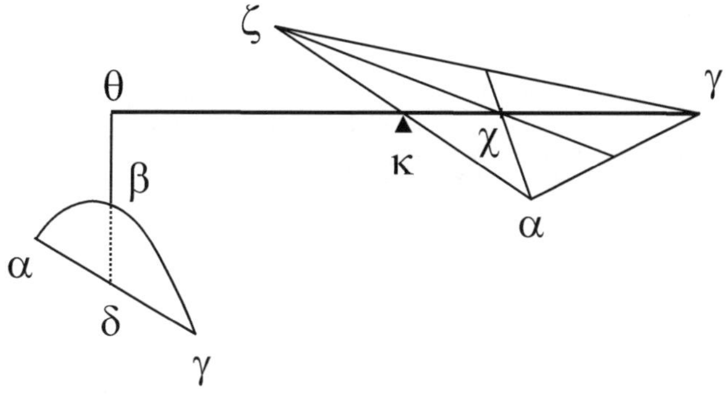

Figure 4.5: This horizontal lever $\theta\gamma$ remains in equilibrium about κ with the segment of parabola $\alpha\beta\gamma$ suspended by a weightless string at θ with its center of gravity vertically below θ, while the triangle $\alpha\zeta\gamma$ is distributed along the arm $\kappa\gamma$. The dashed line is the diameter of the parabola.

The location of the center of gravity of a triangle was first calculated by Archimedes in Propositions 13 and 14 of his work *On the Equilibrium of Planes*.[5] It was also quoted as a lemma of *The Method*, as quoted in Section 4.1.

In the case of the triangle of Figure 4.5 the straight line $\kappa\gamma$ connects the vertex γ to the middle point κ of the opposite side $\zeta\alpha$. The center of gravity of this triangle is located at the point χ of $\kappa\gamma$ dividing it in such a way that

$$\frac{\kappa\gamma}{\kappa\chi} = \frac{3}{1} \,. \tag{4.4}$$

Therefore, the lever will remain in equilibrium in the situation of Figure 4.6.

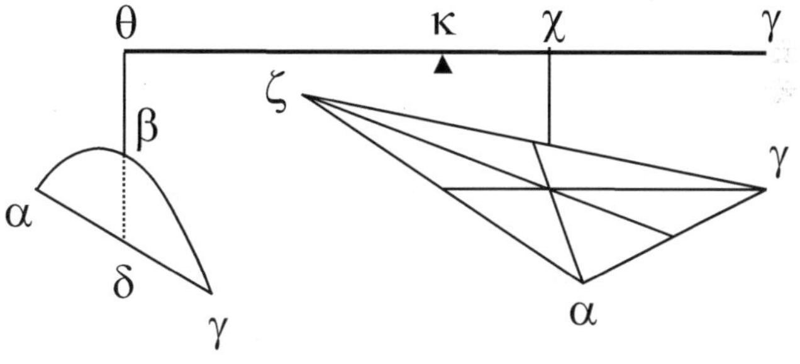

Figure 4.6: The horizontal lever remains in equilibrium about κ with the segment of parabola $\alpha\beta\gamma$ suspended by a weightless string at θ, while the triangle $\alpha\zeta\gamma$ is suspended by a weightless string at χ such that $\kappa\chi = \kappa\gamma/3$.

Utilizing the law of the lever, Equation (2.1), the proportionality between weights and areas, and Equation (4.4), the equilibrium represented by Figure

[5] [6, pp. 198-201], [11, pp. 309-312] and [7, Subsection 10.7.2: Archimedes's Calculation of the CG of a Triangle, pp. 215-217].

4.6 can be written as:

$$\frac{\text{segment of parabola } \alpha\beta\gamma}{\text{area of the triangle } \alpha\zeta\gamma} = \frac{\kappa\chi}{\theta\kappa} = \frac{1}{3} . \tag{4.5}$$

From Figure 4.3 it can be shown that

$$\text{area of the triangle } \alpha\zeta\gamma = 4(\text{area of the triangle } \alpha\beta\gamma) . \tag{4.6}$$

Equations (4.5) and (4.6) yield:

$$\frac{\text{segment of parabola } \alpha\beta\gamma}{\text{area of the triangle } \alpha\beta\gamma} = \frac{4}{3} . \tag{4.7}$$

This is the final result obtained by Archimedes, the quadrature of a parabola. It was obtained combining geometric results with the law of the lever. He expressed it with the following words:[6]

> Any segment of a parabola is four-thirds of the triangle which has the same base and equal height.

4.2.1 Importance of Theorem I

The main aspects of this Theorem are emphasized here.

- Archimedes mentioned in his letter to Eratosthenes that this was "the very first theorem which became known to him by means of mechanics."[7] Therefore, it was no coincidence that he presented this Theorem as the first theorem of *The Method.*

- The geometric proof of this Theorem was known for a long time. This proof is contained in his work *Quadrature of the Parabola.*[8] In this work there is the following statement:

  > [...] I set myself the task of communicating to you [Dositheus], as I had intended to send to Conon, a certain geometrical theorem which had not been investigated before but has now been investigated by me, and which I first discovered by means of mechanics and then exhibited by means of geometry.

  There are two important points to emphasize in connection with this quotation. The first one is that Archimedes was the first scientist to obtain the quadrature of the parabola. Nobody before him had enunciated this result nor provided a demonstration. The second point is that this result was first discovered by means of mechanics. Only after obtaining this result mechanically did he find a geometric proof of the Theorem. With the

[6] [4, p. 14].
[7] [4, p. 14].
[8] [6, Quadrature of the Parabola, pp. 233-252].

discovery of Archimedes's palimpsest this mechanical method was finally revealed. In particular, Archimedes considered a lever in equilibrium under the action of the gravitational attraction of the Earth, with a parabola and a triangle suspended along the arms of this lever at specific distances from the fulcrum, as indicated in Figure 4.6. When the center of gravity of the triangle is known, as given by Equation (4.4), the equilibrium of the lever yields the area of the parabola in terms of the area of the triangle which has the same base and equal height.

- The argument might also be inverted. In the *Quadrature of the Parabola* Archimedes demonstrated geometrically that any segment of a parabola is four-thirds of the triangle which has the same base and equal height.[9] One of his proofs of this result is purely geometrical and does not utilize a lever. Combining this result with the law of the lever, Equation (2.1), and the equilibrium represented by Figure 4.6, yields Equation (4.4), namely, the center of gravity of the triangle. This reversed argument suggests a third procedure to calculate the center of gravity of a triangle, apart from the two earlier procedures presented in his work *On The Equilibrium of Planes*.[10]

- In this work Archimedes obtained the quadrature of the parabola in terms of the area of the triangle which has the same base and equal height. This is a very important result, to obtain the area of a figure bounded by a curved line in terms of the area of a polygon. Archimedes himself had proved a similar result related to the circle, as was seen in Chapter 3.

  With the *Quadrature of the Parabola* he obtained analogously the area of a figure bounded by a curved line in terms of the area of a specific triangle.

4.3 Physical Demonstration of Theorem II: Volume of a Sphere

The method of Archimedes applied to the calculation of the volume of a sphere is now illustrated. In the second Theorem of *The Method* Archimedes proved:[11]

> That the volume of any sphere is four times that of the cone which has its base equal to the greatest circle of the sphere and its height equal to the radius of the sphere, and that the volume of the cylinder which has its base equal to the greatest circle of the sphere and its height equal to the diameter of the sphere is one and a half times that of the sphere.

To arrive at these results he considered Figure 4.7.

[9] [6, Quadrature of the Parabola, pp. 233-252].

[10] [6, pp. 198-201], [11, pp. 309-312] and [7, Subsection 10.7.2: Archimedes's Calculation of

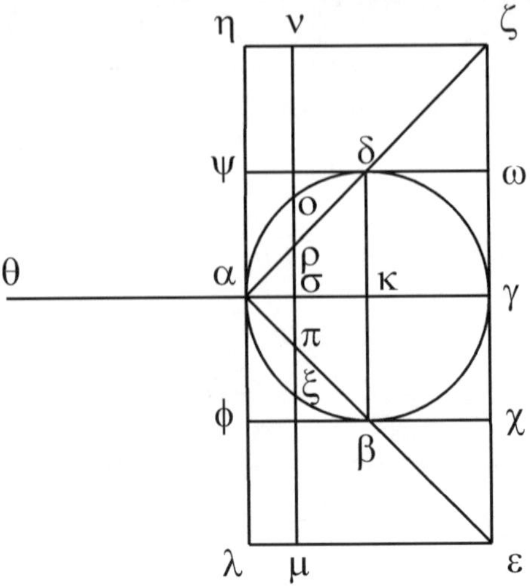

Figure 4.7: Sphere $\alpha\delta\gamma\beta$, cones $\alpha\delta\beta$ and $\alpha\zeta\varepsilon$, and cylinders $\eta\zeta\varepsilon\lambda$ and $\psi\omega\chi\phi$ seen face on.

In Figure 4.7 let $\alpha\delta\gamma\beta$ be the great circle of a sphere with center κ, and $\alpha\gamma$ and $\delta\beta$ diameters at right angles to one another. Let a circle be drawn about $\delta\beta$ as diameter and in a plane perpendicular to $\alpha\gamma$, and on this circle as base let a cone $\alpha\delta\beta$ be described with α as vertex. Let the surface of this cone be produced and then cut by a plane through γ parallel to its base; the section will be a circle on $\zeta\varepsilon$ as diameter. There is then produced a larger cone $\alpha\zeta\varepsilon$ with α as vertex. On the circle $\zeta\varepsilon$ as base let a cylinder $\eta\zeta\varepsilon\lambda$ be erected with height and axis $\alpha\gamma$. Let a circle be drawn about $\omega\chi$ as diameter and in a plane perpendicular to $\alpha\gamma$. On this circle as base let a smaller cylinder $\psi\omega\chi\phi$ be erected with height and axis $\alpha\gamma$. Produce $\gamma\alpha$ to θ, making $\theta\alpha$ equal to $\alpha\gamma$.

Figure 4.8 shows Figure 4.7 in perspective. There are five volumetric bodies, namely, the sphere $\alpha\delta\gamma\beta$, the cones $\alpha\delta\beta$ and $\alpha\zeta\varepsilon$, and the cylinders $\eta\zeta\varepsilon\lambda$ and $\psi\omega\chi\phi$.

Draw any straight line $\nu\mu$ in the plane of the circle $\alpha\delta\gamma\beta$ and parallel to $\delta\beta$, as in Figure 4.7. Let $\nu\mu$ meet the circle at o and ξ, the diameter $\alpha\gamma$ at σ, and the straight lines $\alpha\zeta$ and $\alpha\varepsilon$ at ρ and π, respectively.

Through $\nu\mu$ draw a plane at right angles to $\alpha\gamma$. This plane will cut the large cylinder $\eta\zeta\varepsilon\lambda$ in a circle with diameter $\nu\mu$, the sphere in a circle with diameter $o\xi$, and the large cone $\alpha\zeta\varepsilon$ in a circle with diameter $\rho\pi$. All these circles have a center at σ, Figure 4.9.

From the geometry of Figure 4.7 Archimedes proved that[12]

---

the CG of a Triangle, pp. 215-217].

[11] [3, p. 322].

[12] [4, p. 19].

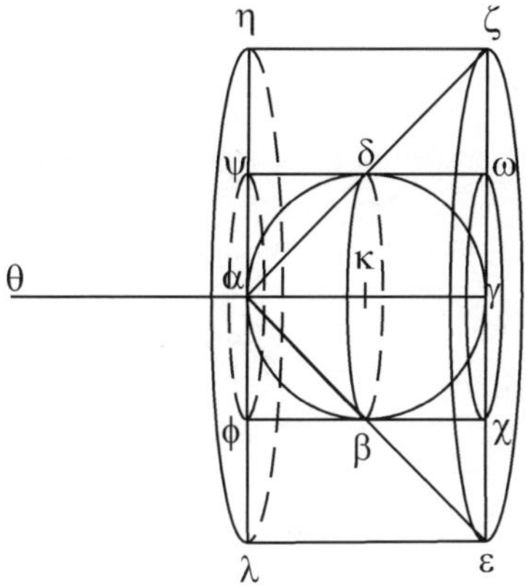

Figure 4.8: Sphere $\alpha\delta\gamma\beta$, cones $\alpha\delta\beta$ and $\alpha\zeta\varepsilon$, and cylinders $\eta\zeta\varepsilon\lambda$ and $\psi\omega\chi\phi$ seen in perspective.

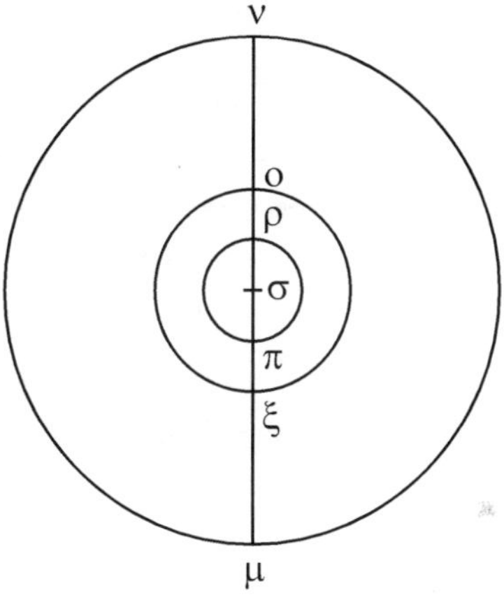

Figure 4.9: The plane through $\nu\mu$ at right angles to $\alpha\gamma$ cuts the large cylinder $\eta\zeta\varepsilon\lambda$ in a circle with diameter $\nu\mu$, the sphere in a circle with diameter $o\xi$, and the large cone $\alpha\zeta\varepsilon$ in a circle with diameter $\rho\pi$. All these circles have the same center σ.

$$\frac{\theta\alpha}{\alpha\sigma} = \frac{\nu\mu \cdot \nu\mu}{o\xi \cdot o\xi + \rho\pi \cdot \rho\pi} . \tag{4.8}$$

Since Eudoxus and Euclid it was known that[13]

> Circles are to one another as the squares on the diameters.

Therefore, Equation (4.8) can be written as:

[13] [13, Proposition 2, Book XII].

$$\frac{\theta\alpha}{\alpha\sigma} = \frac{\text{circle with diameter } \nu\mu}{(\text{circle with diameter } \xi o) + (\text{circle with diameter } \pi\rho)} \, . \tag{4.9}$$

Archimedes then considered the circles with diameters $\nu\mu$, $o\xi$ and $\rho\pi$ as having weights proportional to their areas. Let $\theta\gamma$ be regarded as the bar of a lever, α being its fulcrum, the middle point of $\theta\gamma$. The law of the lever, Equation (2.1), combined with Equation (4.9), imply that this lever will remain in equilibrium if the heavy circular section $\nu\mu$ remains where it is, suspended by its center of gravity σ, while, simultaneously, the heavy circular sections $o\xi$ and $\rho\pi$ are placed at the left extremity of the lever, with their centers of gravity acting at θ. This equilibrium is represented in Figure 4.10.

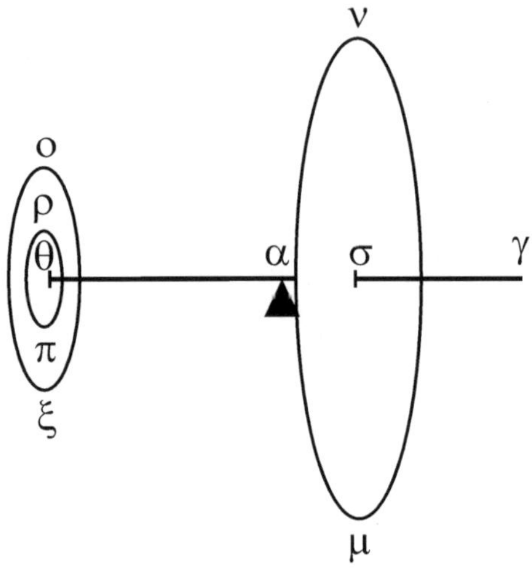

Figure 4.10: The horizontal lever $\theta\gamma$ remains in equilibrium about the fulcrum α when the circle $\nu\mu$ is suspended at σ while the circles $o\xi$ and $\rho\pi$ are simultaneously suspended at θ.

Therefore, the circle $\nu\mu$ in the large cylinder $\eta\zeta\varepsilon\lambda$, placed where it is, centered at σ, is in equilibrium, about α, with the circle $o\xi$ in the sphere together with the circle $\rho\pi$ in the large cone $\alpha\zeta\varepsilon$, if both the latter circles are placed with their centers of gravity at θ.

Figure 4.11 presents the same configuration of equilibrium of Figure 4.10, but now with the two circles at the left suspended by weightless strings. In this configuration the lever remains in equilibrium.

Similarly for the three corresponding circular sections made by a plane perpendicular to $\alpha\gamma$ and passing through any other straight line in the parallelogram $\eta\varepsilon$ of Figure 4.7 parallel to $\zeta\varepsilon$.

By dealing in the same way with all the sets of three circles in which planes perpendicular to $\alpha\gamma$ cut the large cylinder $\eta\zeta\varepsilon\lambda$, the sphere and the large cone

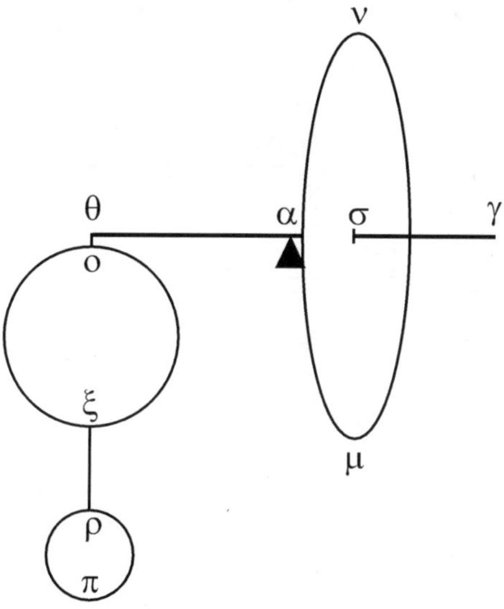

Figure 4.11: The horizontal lever $\theta\gamma$ remains in equilibrium about the fulcrum α when the circle $\nu\mu$ is suspended at σ, while the circles $o\xi$ and $\rho\pi$ are suspended by weightless strings at θ.

$\alpha\zeta\varepsilon$, and which make up these solids respectively, it follows that the large cylinder, so placed, will be in equilibrium about α with the sphere and the large cone, when both are placed with their centers of gravity at θ. This is represented in Figure 4.12 with the cylinder and large cone suspended at θ by weightless strings.

By one of the lemmas of *The Method*, as quoted in Section 4.1, the point κ, the middle point of $\alpha\gamma$, is the center of gravity of the cylinder. Therefore, the cylinder can be suspended at κ by a weightless string without disturbing the equilibrium of the lever, as represented in Figure 4.13.

By the law of the lever, Equation (2.1), together with the proportionality between volumes and weights, the equilibrium of the lever represented in Figure 4.13 can be written as:

$$\frac{\text{large cylinder}_{\eta\zeta\varepsilon\lambda}}{\text{sphere}_{\alpha\delta\gamma\beta} + \text{large cone}_{\alpha\zeta\varepsilon}} = \frac{\theta\alpha}{\alpha\kappa} = \frac{2}{1} \, . \tag{4.10}$$

Proposition 10 of Book XII of *The Elements* of Euclid states that:[14]

> Any cone is a third part of the cylinder which has the same base with it and equal height.

That is, the large cone $\alpha\zeta\varepsilon$ is a third part of the large cylinder $\eta\zeta\varepsilon\lambda$:

$$\text{large cone}_{\alpha\zeta\varepsilon} = \frac{1}{3}(\text{large cylinder}_{\eta\zeta\varepsilon\lambda}) \, . \tag{4.11}$$

[14][13, Proposition 10, Book XII].

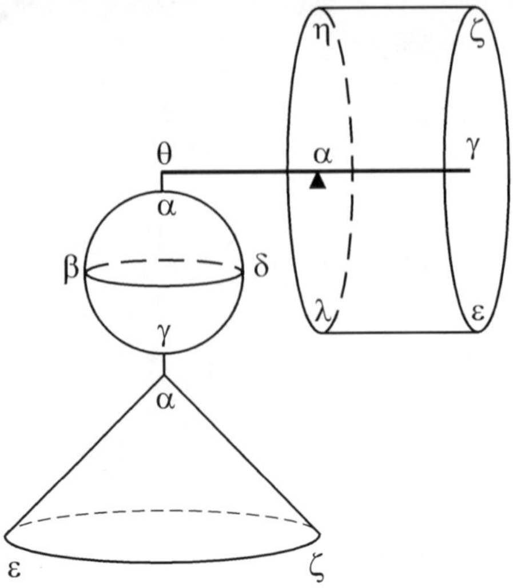

Figure 4.12: This horizontal lever $\theta\gamma$ remains in equilibrium about α with the sphere $\alpha\delta\gamma\beta$ and the large cone $\alpha\zeta\varepsilon$ suspended at θ by weightless strings, while the axis of the large cylinder $\eta\zeta\varepsilon\lambda$ is distributed along the arm $\alpha\gamma$ of the lever.

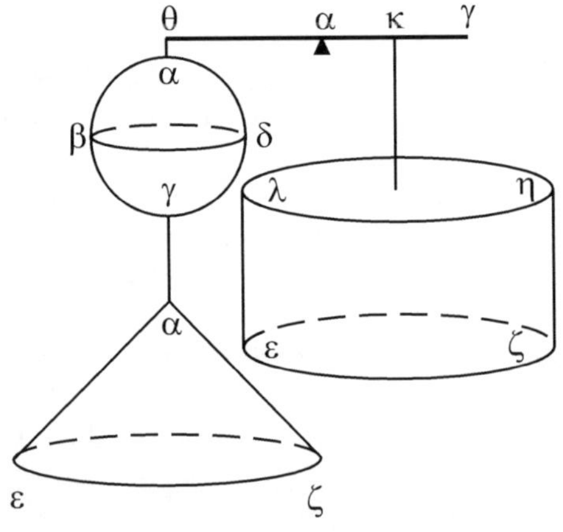

Figure 4.13: The lever of Figure 4.12 remains in equilibrium about α with the sphere and large cones suspended by weightless strings at θ, while the large cylinder is suspended by a weightless string at κ, the middle point of $\alpha\gamma$.

Equations (4.10) and (4.11) yield

$$2(\text{sphere}_{\alpha\delta\gamma\beta}) = \text{large cone}_{\alpha\zeta\varepsilon} \ . \tag{4.12}$$

As the large cone $\alpha\zeta\varepsilon$ has twice the height of the small cone $\alpha\delta\beta$ and its base has twice the diameter of the small cone $\alpha\delta\beta$ this yields:

$$\text{large cone}_{\alpha\zeta\varepsilon} = 8(\text{small cone}_{\alpha\beta\delta}) \ . \tag{4.13}$$

Equations (4.12) and (4.13) yield the first part of this Theorem, namely:

$$\text{sphere}_{\alpha\delta\gamma\beta} = 4(\text{small cone}_{\alpha\beta\delta}) \ . \tag{4.14}$$

Archimedes expressed this result as follows:[15]

> The volume of any sphere is four times that of the cone which has its base equal to the greatest circle of the sphere and its height equal to the radius of the sphere.

Archimedes continued to prove the second part of this Theorem. From Figure 4.8:

$$\text{small cone}_{\alpha\beta\delta} = \frac{1}{3}(\text{cylinder}_{\psi\delta\beta\phi}) = \frac{1}{6}(\text{small cylinder}_{\psi\omega\chi\phi}) \ . \tag{4.15}$$

Equations (4.14) and (4.15) yield

$$\text{small cylinder}_{\psi\omega\chi\phi} = \frac{3}{2}(\text{sphere}_{\alpha\delta\gamma\beta}) \ . \tag{4.16}$$

This completes the proof of the second part of this Theorem.

4.3.1 Importance of Theorem II

There are important points relating to this Theorem.

- The greatest relevance of this Theorem is that for the first time in history Archimedes obtained the volume of a sphere, as was seen in Chapter 3. Let V_S be the volume of a sphere of radius r. This volume is expressed nowadays by the following formula:

$$V_S = \frac{4}{3}\pi r^3 \ . \tag{4.17}$$

- The result of this Theorem was known to the specialists in Archimedes. It appeared in his work *On the Sphere and Cylinder* with a geometric proof.[16] It was only with the discovery of *The Method* that it became apparent how Archimedes originally proved this result. Essentially he utilized a proportion equating the ratio of two distances with another ratio of areas, the law of the lever, and his method of mechanical theorems. He concluded that the lever of Figure 4.13 remains in equilibrium about α with $\alpha\kappa = \theta\alpha/2$. Since Democritus it was known that the volume of a cone is one third the volume of a cylinder with the same base and height.

[15] [11, p. 322].
[16] [6, Proposition 34, pp. 41-44].

This result was first proved rigorously by Eudoxus, being included in *The Elements* of Euclid.[17] Combining this result with the law of the lever and the configuration of equilibrium represented by Figure 4.13, Archimedes could then relate the volume of the sphere with the volume of the cone. Analogously he could relate the volume of the sphere with the volume of the circumscribing cylinder.

After obtaining the volume of the sphere by means of mechanics, he succeeded in obtaining a geometric proof of this Theorem which did not depend upon a lever.

- Nowadays few students are aware that the area of the sphere was also first obtained by Archimedes. The modern formula representing this area is given by:

$$A_S = 4\pi r^2 . \tag{4.18}$$

In Chapter 3 it was seen how Archimedes expressed this result, namely:

> The surface of any sphere is equal to four times the greatest circle in it.

In his work *On the Sphere and Cylinder* this result appears as Theorem 33, while the volume of the sphere appears as Theorem 34. For this reason the commentators of Archimedes thought that initially he had obtained the area of the sphere and only later on did he obtain its volume. It was only with the discovery of *The Method* that this false impression was corrected. Nowadays we know that he first obtained the volume of the sphere utilizing the law of the lever. After obtaining this preliminary result, he was led to the conclusion that the area of the sphere is four times that of the greatest circle in it.

The relevant portion of *The Method* containing this information is quoted here:[18]

> From this theorem, to the effect that a sphere is four times as great as the cone with a great circle of the sphere as base and with height equal to the radius of the sphere, I conceived the notion that the surface of any sphere is four times as great as a great circle in it; for, judging from the fact that any circle is equal to a triangle with base equal to the circumference and height equal to the radius of the circle, I apprehended that, in like manner, any sphere is equal to a cone with base equal to the surface of the sphere and height equal to the radius.

[17] [14, p. 13] and [13, Proposition 10, Book XII].
[18] [4, pp. 20-21].

This can be illustrated by Figure 4.14. In this Figure, a circle is shown with inscribed triangles, and a sphere is represented with inscribed pyramids. To simplify the diagram only three pyramids are shown, but the reader should imagine that the sphere is filled by many pyramids, with their vertexes at the center of the sphere.

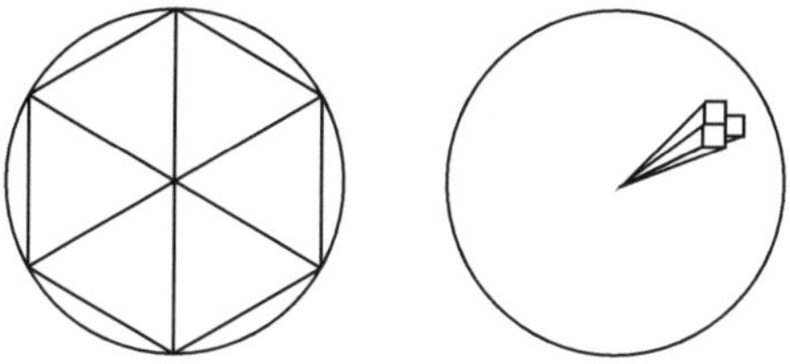

Figure 4.14: A circle with inscribed triangles and a sphere with a inscribed pyramids with their vertexes at the center of the sphere.

That is, when the bases of the inscribed triangles in the circle are decreased and their number are increased, the area of all the triangles will approach the area of the circle. In the limit of infinitely many triangles, the area of the circle will be equal to the area of a triangle with base equal to the circumference and height equal to the radius of the circle, as represented in Figure 3.3. Analogously, when the bases of the inscribed pyramids are decreased and their numbers are increased, the volume of all the cones will approach the volume of the sphere. In the limit of infinitely many pyramids, the volume of the sphere will be equal to the volume of a cone with base equal to the area of the sphere and height equal to the radius of the sphere, as in Figure 4.15.

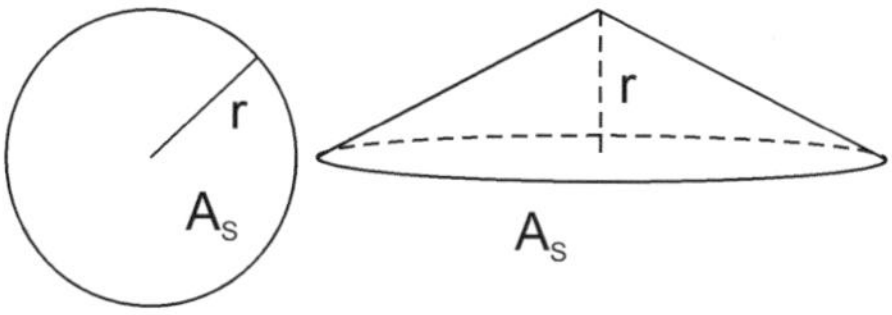

Figure 4.15: The volume of the sphere with radius r and area A_S is equal to the volume of the cone with base A_S and height r.

From the second theorem of *The Method* Archimedes obtained that the volume of any sphere is four times the volume of the cone with base equal to a great circle of the sphere and height equal to its radius. This can be represented by Figure 3.4 and Equation (4.19):

$$V_S = 4V_{Cone} \; . \tag{4.19}$$

The result of Figure 4.15 can be expressed mathematically as follows:

$$V_S = V_{Large\ cone} \ . \tag{4.20}$$

In this Equation the large cone is a cone with height equal to the radius r of the sphere and base equal to the area A_S of the sphere. Equations (4.19) and (4.20) yield

$$V_{Large\ cone} = 4V_{Cone} \ . \tag{4.21}$$

Both cones of Equations (4.20) and (4.21) have the same height r. The area of the base of the large cone is the area A_S of the sphere of radius r, while the base of the small cone is the greatest circle of the sphere, as given by Equation (3.8). These facts show that Equation (4.21) can be written as:

$$A_S = 4A_{Circle} = 4\left(\pi r^2\right) \ . \tag{4.22}$$

That is, the surface of any sphere is four times as great as a great circle in it, as expressed by Archimedes.

- The Theorem relating the volume of the sphere with that of the circumscribing cylinder was considered by Archimedes his greatest achievement. This can be gathered from the fact that he asked his relatives to place upon his tomb a representation of a cylinder circumscribing a sphere within it, together with an inscription giving the ratio which the cylinder bears to the sphere.

  Cicero, when quaestor in Sicily, found the tomb in a neglected state and restored it. It has never been seen since. Cicero wrote the following, as quoted by Rorres:[19]

  > When I was quaestor in Sicily I managed to track down his grave. The Syracusians knew nothing about it, and indeed denied that any such thing existed. But there it was, completely surrounded and hidden by bushes of brambles and thorns. I remembered having heard of some simple lines of verse which had been inscribed on his tomb, referring to a sphere and cylinder modelled in stone on top of the grave. And so I took a good look round all the numerous tombs that stand beside the Agrigentine Gate. Finally I noted a little column just visible above the scrub: it was surmounted by a sphere and a cylinder. I immediately said to the Syracusans, some of whose leading citizens were with me at the time, that I believed this was the very object I had been looking for. Men were sent in with sickles to clear the site, and when a path to the monument had been opened we walked right up to it. And the verses were still visible, though approximately the second half of each line had been worn away.

[19] [15].

4.4 Physical Demonstration of Theorem V: Center of Gravity of a Segment of a Paraboloid of Revolution

This Theorem states the following:[20]

> The center of gravity of a segment of a paraboloid of revolution cut off by a plane at right angles to the axis is on the straight line which is the axis of the segment, and divides the said straight line in such a way that the portion of it adjacent to the vertex is double of the remaining portion.

Figure 4.16 presents the main elements needed to prove this Theorem.

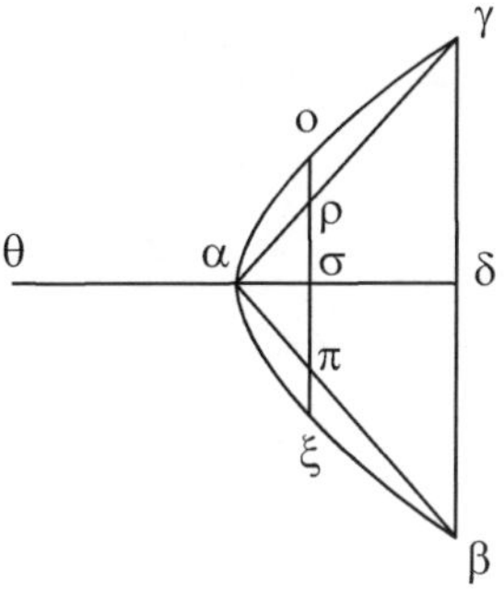

Figure 4.16: Geometric construction of Theorem V with the parabola $\alpha\gamma\beta$.

Let a paraboloid of revolution be cut by a plane through the axis $\alpha\delta$ in the parabola $\alpha\gamma\beta$, Figure 4.16. Let the paraboloid of revolution be cut by another plane at right angles to the axis $\alpha\delta$ and intersecting the former plane in $\gamma\beta$. Produce the axis of the segment $\delta\alpha$ to θ, making

$$\theta\alpha = \alpha\delta \ . \tag{4.23}$$

The base of the segment is the circle on $\gamma\beta$ as diameter. A cone $\alpha\gamma\beta$ has this circle as base and vertex α, so that $\alpha\gamma$ and $\alpha\beta$ are generators of the cone. In the parabola let any double ordinate $o\xi$ be drawn meeting $\alpha\beta$, $\alpha\delta$ and $\alpha\gamma$ in π, σ and ρ, respectively. If now through $o\xi$ a plane be drawn at right angles to $\alpha\delta$, this plane cuts the paraboloid in a circle with diameter $o\xi$ and the cone in a circle with diameter $\rho\pi$.

Figure 4.17 presents Figure 4.16 in perspective.

From the geometry of Figure 4.16 Archimedes proved that[21]

$$\frac{\theta\alpha}{\alpha\sigma} = \frac{\sigma\xi \cdot \sigma\xi}{\sigma\pi \cdot \sigma\pi} \ . \tag{4.24}$$

[20] [4, p. 25].
[21] [4, pp. 26-27].

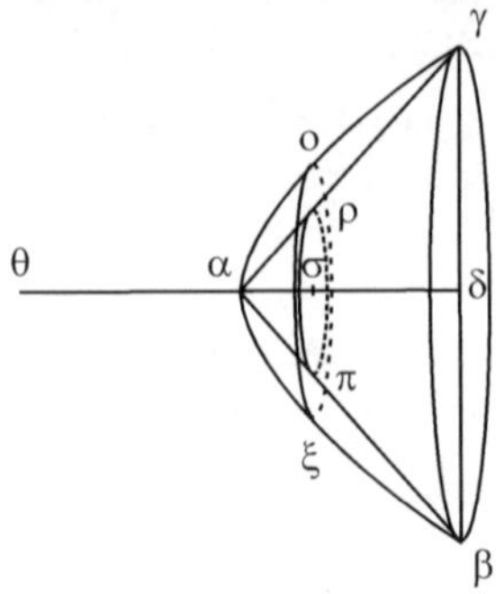

Figure 4.17: Figure 4.16 in perspective.

But circles are to one another as the squares on the diameters. This means that this Equation can also be written as:

$$\frac{\theta\alpha}{\alpha\sigma} = \frac{\text{circle with diameter } o\xi}{\text{circle with diameter } \rho\pi} . \tag{4.25}$$

This is the basic mathematical relation utilized by Archimedes, together with the law of the lever, to prove this Theorem.

Imagine $\theta\delta$ to be the bar of a lever with fulcrum α, its middle point. Suppose weights to be uniformly distributed in the circles, that is, proportional to their areas. Equations (2.1) and (4.25) then lead to a lever in equilibrium. Therefore, the circle $o\xi$ in the paraboloid, in the place σ where it is, remains in equilibrium about α, with the circle $\rho\pi$ in the cone placed with its center of gravity at θ. This configuration is represented in Figure 4.18.

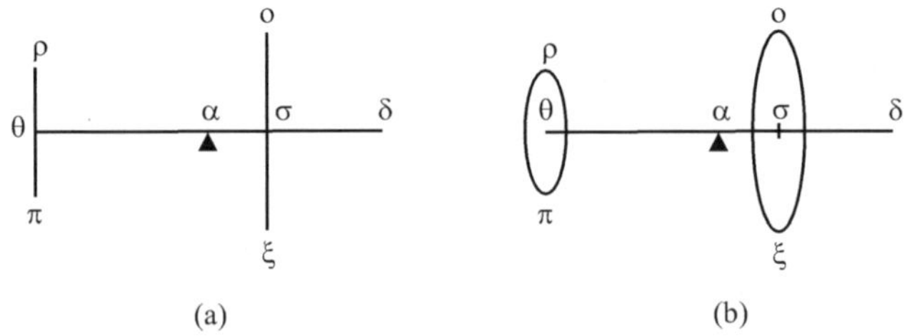

Figure 4.18: Circles in equilibrium upon the lever. (a) Seen face on. (b) Seen in perspective.

Similarly for the two corresponding circular sections made by a plane perpendicular to $\alpha\delta$ and passing through any other ordinate of the parabola. Therefore, when all the circular sections which make up the whole of the segment of the paraboloid and the cone, respectively, are dealt with in the same way, the segment of the paraboloid, so placed, remains in equilibrium about α, with the cone placed with its center of gravity at θ. This is represented in Figure 4.19.

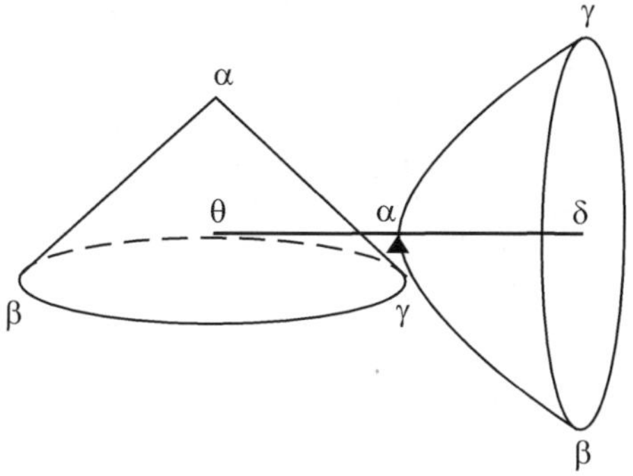

Figure 4.19: Equilibrium of the lever with the segment of the paraboloid distributed over the arm $\alpha\delta$, while the center of gravity of the cone acts only at θ.

Figure 4.20 presents the same configuration of equilibrium with the cone suspended at the extremity θ of the lever by a weightless string, with its center of gravity vertically below θ.

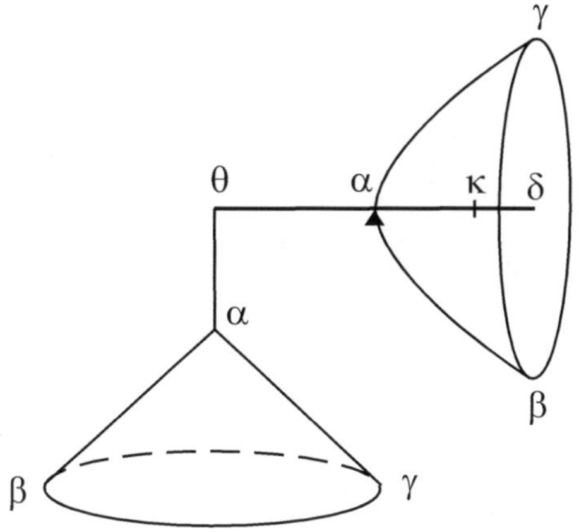

Figure 4.20: Equilibrium of the lever with the segment of the paraboloid distributed over the arm $\alpha\delta$, while the cone is suspended at θ.

By symmetry the center of gravity of the paraboloid is along its axis of symmetry $\alpha\delta$. Let κ be this center of gravity, Figure 4.20. The goal Archimedes set for himself was to find the ratio between $\alpha\kappa$ and $\alpha\delta$. By the sixth postulate of *On the Equilibrium of Planes*, quoted in Subsection 2.1.3, the equilibrium of Figure 4.20 is not disturbed when the paraboloid acts only at κ. This is represented in Figure 4.21 with the paraboloid suspended by a weightless string at κ, with its center of gravity vertically below κ.

The equilibrium represented by Figure 4.21 can be expressed mathematically as:

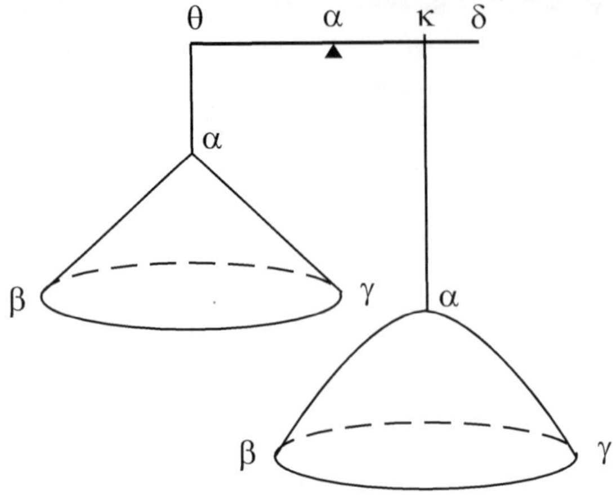

Figure 4.21: Lever in equilibrium about α with the cone acting at θ and the paraboloid acting at κ.

$$\frac{\alpha\kappa}{\alpha\theta} = \frac{\text{cone}}{\text{segment of paraboloid}} . \tag{4.26}$$

In the fourth Theorem of *The Method* Archimedes proved that:[22]

> Any segment of a paraboloid of revolution cut off by a plane at right angles to the axis is one and a half times the cone which has the same base and the same axis as the segment.

This can be expressed mathematically as:

$$\text{segment of paraboloid} = \frac{3}{2}(\text{cone}) . \tag{4.27}$$

Equations (4.26) and (4.27) lead to:

$$\alpha\kappa = \frac{2}{3}(\alpha\theta) = \frac{2}{3}(\alpha\delta) . \tag{4.28}$$

But

$$\alpha\kappa + \kappa\delta = \alpha\delta . \tag{4.29}$$

Therefore, combining Equations (4.28) and (4.29), it can be concluded that the center of gravity of a segment of a paraboloid of revolution is located along its axis of symmetry $\alpha\delta$ in a point κ such that the portion of the axis adjacent to the vertex is double of the remaining portion:

$$\alpha\kappa = 2(\kappa\delta) . \tag{4.30}$$

[22] [4, p. 24].

4.4.1 Importance of Theorem V

Important aspects of this Theorem:

- In Theorem I Archimedes had obtained the unknown area of the parabola utilizing the known area of the triangle, its known center of gravity, and a lever in equilibrium with these two areas suspended by their centers of gravity, Figure 4.6. In Theorem II Archimedes obtained the unknown volume of a solid utilizing the same procedure, as illustrated by Figure 4.13. He knew the volumes of the cylinder and cone, the law of the lever and the ratio of the distances $\alpha\kappa$ and $\theta\alpha$. He was then able to obtain the volume of the sphere in terms of the volumes of the cone and circumscribed cylinder. In Theorem V he obtained for the first time the location of the center of gravity of a body utilizing his method. The configuration of equilibrium which he obtained is that of Figure 4.21. In this case he knew the ratio of the volume of the cone to that of the paraboloid, but did not know the ratio of $\alpha\kappa$ to $\theta\alpha$. He utilized the law of the lever to obtain this unknown ratio, that is, the center of gravity of the paraboloid.

- In his work *On Floating Bodies* Archimedes had stated the precise location of the center of gravity of the paraboloid of revolution.[23] He investigated the different positions of equilibrium in which a right paraboloid of revolution can float in a fluid. But the proof of the center of gravity did not appear in his work *On Floating Bodies*. It was only with the discovery of *The Method* that the procedure used by Archimedes to obtain this center of gravity became known.

  To understand how Archimedes utilized the law of the lever in order to calculate the unknown location of the center of gravity of a body is fascinating.

[23] [6, p. 265] and [11, pp. 380 and 384].

Chapter 5

Conclusion

From what has been seen in this work, the essence of Archimedes's method can be summarized as follows:

1. By geometrical considerations, a proportion is obtained stating the equality of two ratios. One ratio is that of two distances. The other ratio can be that of lengths belonging to certain figures, as in Equation (4.2). The other ratio can also be that of areas belonging to certain figures, as in Equations (4.8) and (4.24).

2. These geometric figures are considered as having weights uniformly distributed. In particular, the weight of each figure will be supposed proportional to its length, area, or volume.

3. These magnitudes are imagined to be suspended upon a lever in equilibrium, according to Equation (2.1). The configurations of equilibrium were represented here by Figures 4.4, 4.10 and 4.18.

4. Each plane figure is considered as being filled up by all straight segments contained in it parallel to a certain direction. Analogously, each solid figure is considered as being filled up by all planes contained in it orthogonal to a certain direction.

5. A lever in equilibrium is then obtained with one or more bodies suspended on one arm of the lever by their centers of gravity, while another body is distributed along the other arm of the lever. These configurations were represented here by Figures 4.5, 4.12 and 4.20.

6. By the crucial sixth postulate of his work *On the Equilibrium of Planes*, quoted in Subsection 2.1.3, he was then able to replace the bodies distributed along one arm of the lever by equal bodies suspended only by their centers of gravity. The sixth postulate guarantees that the levers will remain in equilibrium with these substitutions. These configurations of equilibrium were represented here by Figures 4.6, 4.13 and 4.21.

7. The combination of the law of the lever, Equation (2.1), then yields the area, volume, or center of gravity of a geometric figure when the area, volume, or center of gravity of another figure is known.

Bibliography

[1] D. E. Smith. A newly discovered treatise of Archimedes. *The Monist*, 19:202–230, 1909.

[2] Archimedes. *Geometrical Solutions Derived from Mechanics*. Open Court, Chicago, 1909. German translation from the Greek by J. L. Heiberg. English version translated from the German by L. G. Robinson and reprinted from "The Monist," April, 1909. Introduction by D. E. Smith.

[3] Archimedes. The Method of Mechanical Theorems. In E. J. Dijksterhuis, editor, *Archimedes*, chapter X, pages 313–345. Princeton University Press, Princeton, 1987. Translated by C. Dikshoorn.

[4] Archimedes. The Method of Archimedes. In T. L. Heath, editor, *The Works of Archimedes*, pages 1–51 (Supplement). Dover, New York, 2002. Translated by T. L. Heath.

[5] T. Heath. *A History of Greek Mathematics*. Clarendon Press, Oxford, 1921. Vol. II: From Aristarchus to Diophantus.

[6] Archimedes. *The Works of Archimedes*. Dover, New York, 2002. Translated and edited in modern notation by T. L. Heath.

[7] A. K. T. Assis. *Archimedes, the Center of Gravity, and the First Law of Mechanics: The Law of the Lever*. Apeiron, Montreal, 2nd edition, 2010. ISBN: 9780986492648. Available at: www.ifi.unicamp.br/~assis.

[8] P. Duhem. *The Origins of Statics*, volume 123 of *Boston Studies in the Philosophy of Science*. Kluwer, Dordrecht, 1991. Translated by G. F. Leneaux, V. N. Vagliente and G. H. Wagener.

[9] P. Duhem. *Les Origines de la Statique*, volume 2. Hermann, Paris, 1906.

[10] C. Mugler. *Les Oeuvres d'Archimède*, volume 2: *Des Spirales, De l'Équilibre des Figures Planes, L'Arénaire, La Quadrature de la Parabole*. Budé, Paris, 1971.

[11] E. J. Dijksterhuis. *Archimedes*. Princeton University Press, Princeton, 1987. Translated by C. Dikshoorn.

[12] W. Stein. Der Begriff des Schwerpunktes bei Archimedes. *Quellen und Studien zur Geschichte der Mathematik, Physik und Astronomie. Abt. B: Quellen. I*, pages 221–244, 1930.

[13] Euclid. *The Thirteen Books of The Elements*, volume 3, Books X-XIII. Dover, New York, 1956. Translated with introduction and commentary by Sir Thomas L. Heath.

[14] Archimedes. The Method of Archimedes. In T. L. Heath, editor, *The Works of Archimedes*, pages 1–51 (Supplement). Dover, New York, 2002. Traduzido por T. L. Heath.

[15] C. Rorres. *Archimedes.* Site created and maintained by C. Rorres. Available at: www.math.nyu.edu/~crorres/Archimedes/contents.html.

Made in the USA
Monee, IL
04 November 2026

67375689R00026